Camila Tainá dos Santos Rocha
Maria Viviane Palmeira da Costa
Joelma Pereira da Silva

The Socio-Economic Importance of Banana Culture

Camila Tainá dos Santos Rocha
Maria Viviane Palmeira da Costa
Joelma Pereira da Silva

The Socio-Economic Importance of Banana Culture

for the Municipality of Missão Velha-CE

ScienciaScripts

Cover image: www.ingimage.com

This book is a translation from the original published under ISBN 978-613-9-68946-0.

Publisher:
Sciencia Scripts
is a trademark of
Dodo Books Indian Ocean Ltd. and OmniScriptum S.R.L publishing group

120 High Road, East Finchley, London, N2 9ED, United Kingdom
Str. Armeneasca 28/1, office 1, Chisinau MD-2012, Republic of Moldova, Europe
Printed at: see last page
ISBN: 978-620-8-24144-5

SUMMARY

I dedicate this work to my parents, sisters and all my family who, with great affection and support, have spared no effort to get me to this stage in my life, and to my grandfather, Vicente Santos Monte (in memory).

1 INTRODUCTION

Fruit growing has grown a lot in recent years and has achieved a prominent place on the national scene, both in terms of fresh fruit production and in the industrialization of juices, pulp and marketing, exports, job creation and income. It is an activity of great importance to the population both economically and socially, as it is one of the foods that is most present in our daily lives, as well as being a source of income for family farming, serving the various social classes and providing greater nutritional value for human health.

Fruit growing is one of the branches of agriculture that aims to produce fruit in general in an economical and rational way, with the aim of marketing it. It is an activity of great importance in terms of economic and social aspects and because it represents an important source of vitamins and nutrients (FERREIRA et al., 2014).

With the development of fruit production, it is one of the main income-generating and expanding practices in national agribusiness, and one of the most labor-intensive activities, be it planting seedlings, handling, harvesting or processing. This generates more job opportunities and provides greater profitability for the producing regions. By bringing innovations to small rural producers with the use of irrigation systems that were previously rarely used, today both small producers and small farms can grow good quality fruit all year round.

The banana, Musa spp., is one of the most widely consumed fruits in the world and is grown in most tropical countries. It is an important source of food and can be eaten green or ripe, raw or processed (boiled, fried, baked and industrialized). The fruit contains vitamins (A, B and C), minerals (Ca, K and Fe) and is low in calories (90 to 120 kcal 100 g-1) and fat. It contains approximately 70% water, with the solid material consisting mainly of carbohydrates (23 to 32 g 100 g-1), proteins (1.0 to 1.3 g 100 g-1) and fats (0.37 to 0.48 g 100 g-1) (BORGES et al., 2009).

Bananas are one of the most important fruits in the world and are considered a complementary food in the diet of most low-income populations. It also stands out for being a major generator of income and jobs for those who survive on this branch of fruit growing.

In this way, we highlight the municipality of Missâo Velha, which has been growing a lot with its banana production and in recent years, through this growth, there has been a need to learn more about the socio-economic importance of this crop for the development of the region. Visits were therefore made to producers, associations and regional companies.

Missâo Velha is a Brazilian municipality in the state of Cearâ. Located in the Cariri region, it lies at a latitude of 07°14'59" south and a longitude of 39°08'35" west, at an altitude of 360 meters. Its estimated population in 2016 was 35,409 inhabitants (IPECE, 2017).

2 OBJECTIVES

2.1 General objective

Analyze the socio-economic importance of banana cultivation for the Municipality of Missao Velha.

2.2 Specific objectives

J Identify the benefits of banana production for the municipality;

J Describe the characteristics of the banana crop;

J Show the most commercialized varieties in the region;

3. THEORETICAL FRAMEWORK

3.1 The banana tree

The Musa banana belongs to the Musaceae family and originated on the Asian continent, with other secondary centers of origin in Africa and the Pacific islands. One of the main physiological characteristics of the banana tree is the presence of the "false" trunk, formed by the sheaths of the overlapping leaves. The stem of the banana tree is actually an underground rhizome, and the "fruits", the bananas, are pseudobagas (SENA, 2011).

Silva (et al., 2017) describes the banana tree by citing various authors:

> Banana trees are very sensitive to the competition imposed by weeds for production factors such as light, water and especially nutrients, which can result in a reduction in vigor and a drop in productivity (GOMES et al., 2010; MOURA FILHO et al., 2015). The interference caused by weeds in banana orchards is most critical in the first five months after planting the crop, since at this stage the plants are smaller (LANZA et al., 2017). After this initial period, the banana trees are more vigorous and less sensitive to interference, already producing shade by closing the rows, which prevents or delays the development of weeds (LANZA et al., 2017).

Integrated control is the combination of methods that efficiently combat weeds in banana growing, reducing costs and the use of herbicides; enabling a more ecological environment in the banana plantation; improving and preserving natural resources such as soil and water, providing greater competitiveness and sustainability for the producer, but above all, helping to control banana pests and diseases (CARVALHO, 2000).

Quoting various authors, Santos (2014) describes the banana tree as follows:

> Bananas are the 4th most produced product on the planet. The Series of Proposals for action in production chains of the Bank of Brazil (2010, p. 02). Brazil is characterized by large-scale fruit production, and banana production ranks third (FAO, 2009). In the last three decades, bananas have seen a significant increase (122%) in the volume produced worldwide (SANTOS, 2014).

3.2 Planting the banana tree

When it comes to planting any type of crop, you have to bear in mind the consumer market, the area in which it will be grown and, among these aspects, follow the normal methods that every producer uses before planting any crop, such as clearing the area, plowing, digging holes or holes for planting and setting up the irrigation system.

It is recommended that soil analysis be carried out before planting, as it is through this that we can get a diagnosis of the area to be planted, and if there is a need to fertilize and liming in advance, the better off the producer will be.

According to Borges (2009), the best temperature for the normal development of commercial banana plants is around 28°C, considering the temperature range of 15°C to 35°C as the extreme limits for the rational exploitation of the crop. If there is a supply of water and nutrients, this temperature range induces maximum plant growth.

Bananas are grown in tropical climates. In very cold regions, they do not grow well. They are most commonly grown in the northeast because of the favorable conditions for planting them. On average, banana trees need 1200 to 1800 mm/year of well-distributed rainfall. Wind is one of the factors that can damage banana plantations, causing minor damage and even destroying the entire area.

The preferred soils are those rich in organic matter, well-drained, clayey or mixed, with good water availability and favorable topography. Sandy soils, in addition to low fertility and poor moisture retention, favor the spread of nematodes and should therefore receive more attention (SIMÔES, 2011).

Today, with the expansion of banana cultivation, even sandy soils are being cultivated with banana trees, requiring only more intense fertilization and a greater amount of water for irrigation; because they are nutrient-poor soils, but deep, thus providing greater development of the plant's root system.

The most favorable time to plant banana trees is at the end of the rainy season, with mild rains, as water requirements are lower until three months after planting.

The spacing of any crop is dependent on a number of factors such as variety, soil fertility, the destination of the produce and the movement of machinery to transport the fruit. The most common spacings used for banana trees are rectangular, triangular, square, hexagonal and double row.

In general, planting in a double row is more recommended because varieties of various sizes are used - low, medium and high. Using this spacing makes it easier to inspect the banana trees, helps to carry out plant health treatments and allows for intercropping with other crops.

Banana plant propagation is done asexually, so depending on the grower's intentions or availability of investment, there are various techniques available for producing seedlings, ranging from buds obtained from the rhizome of the mother plant to the use of micropropagated seedlings. Seedlings obtained by micropropagation are preferable because they are generally genetically uniform, healthier and more vigorous. Once the banana plantation has been planted and the appropriate management practices have been adopted, its renewal is guaranteed by the formation of shoots at the base of the plant; this characteristic guarantees the stability of the plantation (SENA, 2011).

Seedlings are one of the most important items when setting up an orchard, be it of any plant species. In banana growing, in particular, the quality of the seedlings depends to a large extent on the success of the venture. As well as having a direct influence on the development and production of the banana plantation, especially in its first cycle, seedlings play a fundamental role in its phytosanitary quality, since a series of diseases and pests carried on the seedling can totally compromise the success of the planting.

The health quality of banana seedlings is one of the most important factors in the longevity and productivity of a banana plantation. The characteristics of the mother plants are usually reflected. Ideally, seedlings should come from nurseries, which are areas established for the sole purpose of producing good quality propagating material. They should be obtained from vigorous plants in optimum phytosanitary conditions and no older than 4 years

(NACIMENTO, 2010).

Banana seedlings should be purchased from good nurseries and obtained from areas established for the sole purpose of production. Banana seedlings are classified as horn, little horn, umbrella, piece of rhizome.

3.4 Variety

In the Brazilian market, the most important cultivars are Cavendish (a group that includes Nanica, Nanicao and Grande Naine), Prata, Maça and Ouro. Other varieties also found with some frequency are Prata-Anâ, Pacovan, Branca and da Terra (SEBRAE, 2008).

Nanica - Also known as green peel, it is a small plant, its bunch weighs between 15 and 40 kg, the leaves are dark green on the upper side.

Nanicao - A medium-sized plant, it is more suitable for export due to its preservation during transportation. The bunch has a cylindrical appearance and weighs between 15 and 45 kg, as well as being very productive.

Prata - Also known as Prata-comum, it is tall. The bunches weigh from 9 to 12 kg and have a pleasant bittersweet taste. Its leaves are very long, providing a larger shaded area.

Pacovan - This is currently the most widely planted cultivar in the north and northeast of the country. It has a tall size, the bunches are conical, with an average weight of 16 kg, the fruit is large, with prominent corners and a thick skin, and has a less intense flavor than the Prata.

Prata-ana - Medium to low-growing cultivar. The bunches weigh from 14 to

16 kg and have an average of 7.6 pencas. The fruit weighs an average of 110g and has a flavor close to that of the Prata cultivar. It is the most widely grown banana variety in the Municipality of Missâo Velha.

Maçâ - Medium-sized (4 m) with bunches averaging around 11 kg and 15 pencas per bunch. The fruit weighs an average of 115 grams and has a white pulp, mildly perfumed and with a pleasant taste.

Terra - It is widely consumed in the Northeast, preferably boiled, fried and/or roasted. It has a tall size, the bunches weigh an average of 25 kg, its flesh is yellow-orange in color and has a "locked" taste, due to its high starch content, even when ripe.

3.5 Main banana diseases

Black Sigatoka *(Mycosphaerella fijiensis):*

The symptoms of this disease appear on the underside of the leaves, in the form of yellow dots that quickly develop into brown streaks, and in more advanced stages the leaves die prematurely.

The most effective method for controlling this disease is chemical, but there are other methods such as pruning the attacked leaves.

Figure 1 Leaf with black Sigatoka

Yellow Sigatoka *(Mycosphaerella musicola):*

It is characterized by the appearance of small spots between the veins of the youngest leaves, which turn into streaks. These streaks turn into necrotic spots that gradually develop into dark lesions.

It often occurs during rainy periods and at high temperatures. Producers are advised to control the disease more rigorously by removing attacked leaves and draining the soil. It is also important to choose a more resistant variety.

Figure 2- Leaf with yellow Sigatoka-

Source: Agronoticias

Moko or bacterial wilt *(Ralstonia solanacearum):*

The symptoms of this disease are very similar to those of Panama disease, but the difference is that the fruit is characterized by the disease, which stands out for its early and irregular ripening within the bunch, with dry and black rot in the flesh.

It is controlled by eliminating diseased plants, i.e. eliminating the terminal part of the ricotta by breaking or cutting it off after the pods have formed, and by planting healthy seedlings.

Figure 3- Fruits with Moko

Source: Adapar

Panama disease *(Fusarium oxysporum f. sp. Cubense):*

It occurs in the pseudostem, causing obstruction of the nutrient-conducting vessels, which is characterized by yellowing of the leaves.

younger leaves to the older ones, then the leaves wither, dry and break.

The main way this disease spreads is through contact between the roots of healthy plants and diseased plants. In order to control this disease, it is recommended to carry out an inspection at the time of planting, thus eliminating seedlings that show symptoms or suspicion of the disease, as well as avoiding areas that have already had an incidence of the disease.

Figure 4- Banana tree with Panama disease

Source: Embrapa

3.6 Main banana pests

Banana rhizome borer *(Cosmopolites sordidus):*

It is considered to be a severe pest because of the damage it causes to banana trees. In the larval stage, it causes the greatest damage to the plant by building small galleries inside the rhizome, causing the leaves to turn yellow and dry out. This pest can be controlled by using healthy plants or by cleaning up the banana plantation and cutting off parts of the roots affected by the pest.

Figure 5- Rhizome borer adult specimen

Source: Cultivar

Cavernicola nematode (*Radopholus similis)*

These are very small worms that penetrate the roots, causing damage and impairing the absorption of water and nutrients from the soil, subsequently causing the plant to weaken and die.

This nematode attacks the roots and rhizome causing extensive necrosis, which leaves the plant very vulnerable to toppling by the wind or by the weight of the bunch itself due to the lack of sustainability of the plant by the roots. This pest is controlled by cleaning the seedlings, cutting and removing all the necrotic roots and tissues.

There are several types of nematode, of which the cavernicola (*Radopholus similis*) is the most important due to its damage.

Figure 6 - Plant killed by nematodes

Source: Fitocon - Consulting

3.7 Management

Management is a way of preventing the plant and the soil from suffering damage in the future, i.e. actions designed to increase productivity and possible degradation of the environment, which will help the producer to control and eliminate pests and diseases.

Proper management is very important for the development of the crop in the field, as it is dependent on various factors such as: pest control, irrigation system, soil preparation, correct fertilization, machinery control and harvest time.

Cleaning is a cultural treatment that consists of eliminating hanging and dry mature leaves, thus enabling greater luminosity and control of pests and diseases, with the aim of providing an improvement in the growth and development phase, improving uniformity, size, shape and color within the standards and requirements of the consumer market, in addition to maintaining greater phytosanitary control of the banana plantation.

3.8 Harvesting

Harvesting is a very important time in banana production, as it is largely responsible for the quality and appearance of the final product, i.e. the fruit. We have to keep the consumer market in mind, as it dictates the rules of the game, in the case of supply and demand for production, whether for fresh consumption, export and/or industrialization.

Pre-harvesting is a practice of the utmost importance for the quality of the fruit, i.e. it is related to the taste, aroma, size, internal and external appearance of the fruit, exerting a great influence on its preservation, through this practice we can have greater control of the crop.

According to Nascimento (2009):

> The bunches must be transported to the place where they will be dropped and packed, either manually (on the shoulder) or mechanically, in car bodies or tractor trailers, lined with synthetic foam. Special care must be taken to avoid bruising the fruit, which has a thin skin. The bunches must be hung up to be dropped. The fruit should be removed from the bunch into a tank filled with water and detergent, where they will be washed, divided into bouquets and then packed in cardboard, wooden or plastic boxes made specifically for fruit. In all cases, the dimensions are 52 x 39 x 24.5 cm (length x width x height), with a capacity for approximately 18 kg of fruit. The boxes should be lined with low-density plastic to protect the fruit from abrasions.

3.9 Irrigation system

Spot irrigation consists of applying water directly to the plant's root system, forming a wet bulb; it is subdivided into drip and micro-sprinkler irrigation, and its wetted area varies from 20 to 80% of the total area, which can result in greater efficiency and water savings. One of the advantages of this system is that evaporation losses are much lower when compared to other types of irrigation.

The most widely used irrigation system in banana growing is micro-sprinkling, which consists of simulating a small shower of rain by pouring water directly onto the plant's root system. This system is most used by banana growers because it makes it possible to irrigate more than two plants in less time, as well as allowing the use of fertigation, which consists of applying fertilizers and some pesticides via irrigation water.

The initial cost is relatively high, and it is recommended for crops with high economic value and greater spacing between rows of plants. It is a method that allows a high degree of automation, which requires less labor in the operation (GUEDES, 2016).

4. THE SOCIO-ECONOMIC IMPORTANCE OF BANANAS

Bananas are very important due to the amount of nutrients in their fruit. As well as being very tasty, they are rich in potassium, which helps the human body function.

Bananas are one of the most important fruits in the world, both in terms of production and marketing. For many countries, in addition to being a complementary food in the population's diet, bananas are of great social and economic importance, serving as a source of income for many farming families, generating jobs in the countryside and in the city and contributing to the development of the regions involved in their production. For other countries, it is an export product responsible for a very significant part of the income from agricultural exports (FIORAVANÇO, 2003).

4.1 Job creation

Banana cultivation has been growing a lot in the northeast, especially in the states of Bahia, Ceará, Sergipe and others. As well as keeping people in the countryside, it brings about a natural agrarian reform, as it is exploited by small farmers on farms, practically using family labor.

Fruit growing stands out as an important generator of formal jobs in the northeastern rural sector. Benefiting from a good supply of labor, environmental conditions and constitutional funding, the permanent crop fruit segment saw average annual growth of 4.11% p.a. in the number of direct formal jobs in the period from 2010 to 2015, despite the drought (VIDAL, 2016).

4.2 Family farming

Family farming is understood as the set of activities carried out on small properties by groups of families in which their production is destined for their own consumption and the surplus is marketed to meet personal needs.

Family farming is one of the most evident political forces of the last 20 years in the various countries of Latin America. The social advances made in the countryside during this period are largely the result of public policies that were born out of the demands made by the women and men who live off the land. A significant part of the reduction in poverty in the countries of the continent, for example, is due to the rise of a new rural middle class. In

Brazil alone, this section of society, which was 13.6% of the country's population in 1992, jumped to 20.6% in 2003 and reached 35.4% in 2009. This growth is faster than that of the middle class as a whole (OTERO, 2015, p. 17).

Family farming has the unique potential to have an impact on food supply, generate income, reduce poverty and increase the productive and economic inclusion of rural families (WOLF; ARIAIS, 2015, p. 21).

4.3 Banana production in Brazil

Banana is one of the most widely consumed fruits in Brazil and is a staple of the diet of all social classes. Today it is grown in all tropical regions and exported all over the world.

According to the IBGE, Brazil's banana production in 2014 was 7.1 million tons, 2.7% more than in the previous year. The fruit is grown in all Brazilian states and is an important source of nutrients, especially for rural populations further away from urban conglomerates (LSPA, 2014).

4.4 Marketing

Most of the fruit produced in the northeast is consumed domestically, with only a small portion being exported. Today, bananas are considered an ideal investment, as long as investors produce in favorable conditions for their commercialization and development, since it is a crop that produces all year round depending on soil and climate conditions.

This crop can be marketed locally, through middlemen, at open-air fairs, included in social programs such as school meals, hospitals, among others. Because it is a very tasty fruit, rich in nutrients and vitamins, it is a great source of energy, being consumed by all social classes, with an affordable price making it easy to sell.

5. METHODOLOGY

The research carried out is exploratory in nature, with the aim of discovering new knowledge about the subject, and is characterized by a more informal approach, flexibility and creativity during the dialogue.

"Exploratory research carries out precise descriptions of the situation and aims to discover the relationships that exist between its component elements" (CERVO; BERVIAN, 2002, p. 69).

The main research tool was a semi-structured interview applied to the farmers, which followed a script in which they were free to ask new questions during the interview (APPENDIX A).

The interview is a meeting between two people in order for one of them to obtain information on a particular subject through a professional conversation. It is a procedure used in social research to collect data or to help diagnose or treat a social problem (MARCONI; LAKATOS, 2006, p. 92).

The data for the development and conclusion of this research was obtained through an interview with banana producers in Missao Velha. Thus, on April 4, 2018, a visit was made to the properties of producers Diego and Gilson in the city of Missao Velha, and later, on April 12, 2018, another application was made in the district of Missao Nova at the Sitio da Serra company, where a conversation was held with Edinaldo, the company's coordinator. And the last visit was to the Sitio Barreiras Banana Producers' Association on April 26, 2018, with the aim of gaining knowledge about the socio-economic importance of the crop for the municipality.

The following research was carried out in the municipality of Missao Velha, in the following communities: Missao Nova, Sitio terra nova, Sitio Barreiras and Sitio Chico Gomes in Jamacaru.

The interview script was essential for achieving the expected results, so information was collected from the respective properties mentioned above on the planting season, varieties, production costs, marketing, irrigation system used, diseases, fertilization, total area cultivated and also the social and economic aspects of this crop in the life of each producer.

5.1 Banana producers in Missao Velha

Producer I

The first interview was carried out at the terra nova site, which has been planting bananas for two years. The area under cultivation is seven hectares of the prata anâ variety, planted manually with a 5m x 5m spacing, the seedlings used to plant the bananas were bought from a seedling propagation laboratory, which at the time cost 1.00 a plant, the system used is by micro-sprinklers, irrigation is carried out in the morning and evening, the application time is 1.5 hours in each sector.

Fertilization is done manually, using fertigation, which consists of applying fertilizers via irrigation water. Harvesting is done manually with the help of three employees, 30,000 to 40,000 bananas are harvested each week, and the fruit is sold locally.

Producer II

He has been growing bananas for more than 20 years. The varieties produced on his property are Prata Anâ and Pacovan, which are resistant to nematodes, and moderately resistant to the Rhizome Borer, they are susceptible to Yellow Sigatoka disease, and moderately resistant to Panama disease.

The seedlings used for planting were obtained through donations. These two varieties are planted on one hectare, on which no irrigation system is used, only rainwater, harvesting is done manually, the annual production is 24 million and the market is sold through a middleman.

5.2 Sitio da Serra Company

The research was carried out at the Sitio da Serra branch in Missâo Nova. The company has been in business since 2007 and is based in the municipality of

Barbalha. It came about through a project that his father brought to Sitio Barreiras, in which he involved his sons and today it is one of the largest producers in the Cariri region along with his brother.

Unit two, as it is called, has approximately 60 hectares of land, 44 hectares of which are planted with bananas of the Prata X variety, which is ready to be sold just 16 weeks after it is due to be harvested, and the Willian (nanica) variety, which is more resistant to Panama disease. We are also now starting to grow Formosa and Hawaii papaya.

The seedlings obtained for planting the banana plantation were propagated in the laboratory, which are produced through the cloning process (micropropagation) and acclimatized in greenhouses, thus choosing only quality seedlings, free from pests and diseases, and adapted to the climatic conditions of the region.

The irrigation system used to irrigate the banana plantation is micro-sprinkling, which consists of water droplets similar to a softer, more uniform rain.

The bananas are divided into plots to make it easier to control production, so only the bunch that is ready to be harvested is taken out of the field. Harvesting is done with the help of an aerial cable that transports the fruit to the shed, where the banana processing and quality control process begins.

The bananas are marketed in this way: the company produces them and the barrier site buys them. The company also takes part in projects like mesa Brasil and donates fruit to hospitals.

5.3 Sitio Barreiras Banana Producers Association

The last interview was carried out at the Association of Banana Producers in Sitio Barreras, which was founded in 2008 and has 32 members, of which only four were interviewed. Producer I

He has been growing bananas for 15 years. Before he started growing fruit, he planted peanuts, a legume rich in nutrients, but at the time there was not enough return on investment.

In view of this, he looked for a crop that would provide a quick return and would adapt to the region's climatic conditions, so he planted bananas on his property because it was a differentiated income option, due to market acceptance and the fact that the crop produces all year round under irrigation conditions and proper management.

The variety produced on the property is Prata ana, with 5,000 trees planted. The seedlings were obtained from Sitio Barreiras. The system used for irrigation is micro-sprinkler irrigation, which is highly profitable and is a characteristic of the localized irrigation method, which applies water and nutrients in a larger fraction and volume in less time, thus exploiting the soil and the roots of the plant. The fruit is marketed by middlemen.

Producer II

He has been growing bananas for 12 years. The variety grown on his property is prata anâ, with six hectares planted. One of the advantages of banana growing is the faster return on investment compared to other fruit crops.

Its production reaches 380 million a year, and it is marketed through middlemen. Nowadays, this is a technique used by many small producers who don't have transportation to take their produce to the market.

Producer III

He has been growing bananas for eight years, started his production through the association, and is now very satisfied with his production. There are eight hectares planted with the prata anâ variety, and his production is irrigated by micro-sprinklers in adequate quantities and proportions, thus promoting a larger wetted area, allowing for greater uniformity of planting.

Through the association he has acquired new knowledge that has contributed to better management, improving his production and fruit quality. His yield reaches 400,000 per year, and the fruit is sold through a middleman.

Producer IV

According to the producer, before he started investing in banana farming he grew corn, beans and peanuts. He changed because he felt the need to improve his living conditions, so he saw bananas as a differentiated income option, due to market acceptance and the fact that the crop produces all year round under conditions of irrigation and proper management.

He has been growing bananas for four years now, and is one of the newest members of the association, with four hectares of the prata anâ variety grown on his property, with an annual production of 300 thousand. Today he is very satisfied with his production. Since he started growing bananas, his life and that of his family has improved a lot.

The field research lasted three days (April 4, 12 and 26, 2018), during which interviews were conducted in the city of Missâo Velha, depending on the availability of the producers. The interviews took place in the morning, afternoon and evening, requiring an average of two hours.

At the end of the survey, a total of seven people were interviewed, divided into two producers, four associates and one company, closing the cycle of producer referrals. The results were then analyzed and graphs were drawn up to show the results more clearly.

6. RESULTS AND DISCUSSION

As can be seen in the graph, bananas are indeed very important for the municipality of Missao Velha, both in social and economic terms for the region, as well as being a major generator of jobs and income for the population that survives on growing this crop.

Graph 1 - Perception of the importance of bananas for the producers surveyed

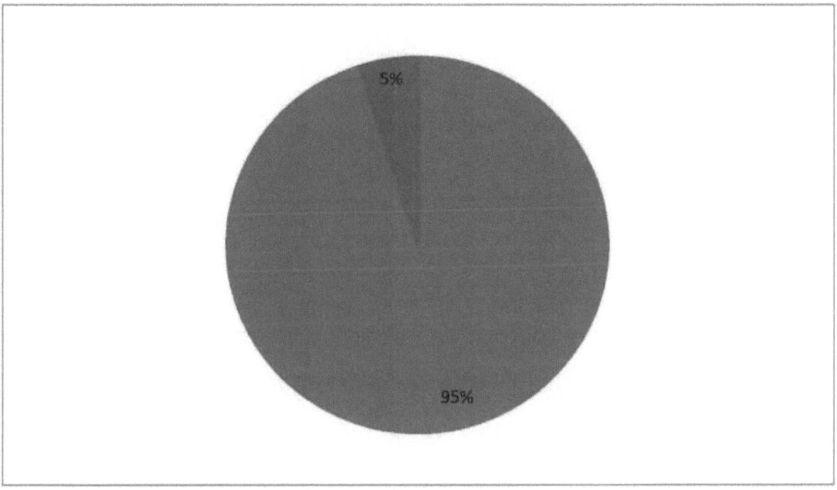

Source: ROCHA, 2018

In the Northeast, it is generally exploited by small rural producers who use family labor, whether in cultivation, harvesting, handling, marketing, etc., thus helping to keep people in the countryside; this cultivation is done on small properties, which causes a natural agrarian reform.

Graph 2 - Varieties grown by the farmers surveyed

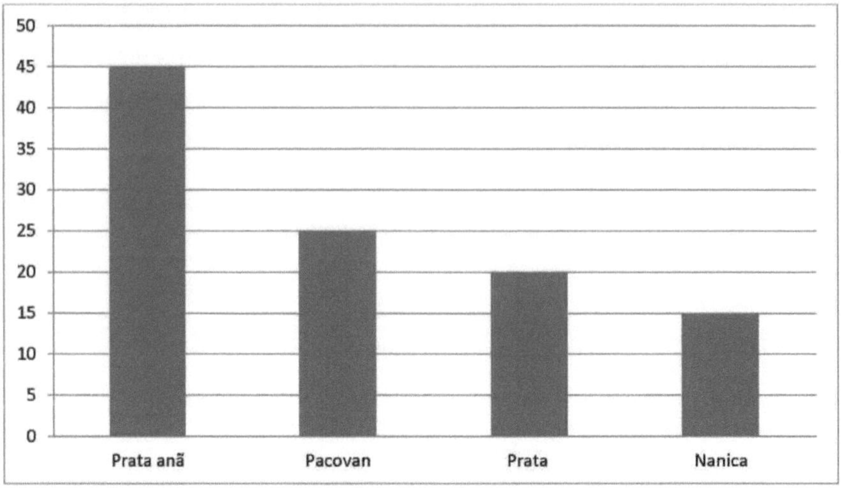

Source: ROCHA, 2018

The most widely cultivated varieties in the region are prata anâ and pacovan because they are more productive and therefore more profitable. It is worth noting that the common prata has a higher commercial value due to its flavor, although its productivity is low.

Graph 3 - Banana marketing

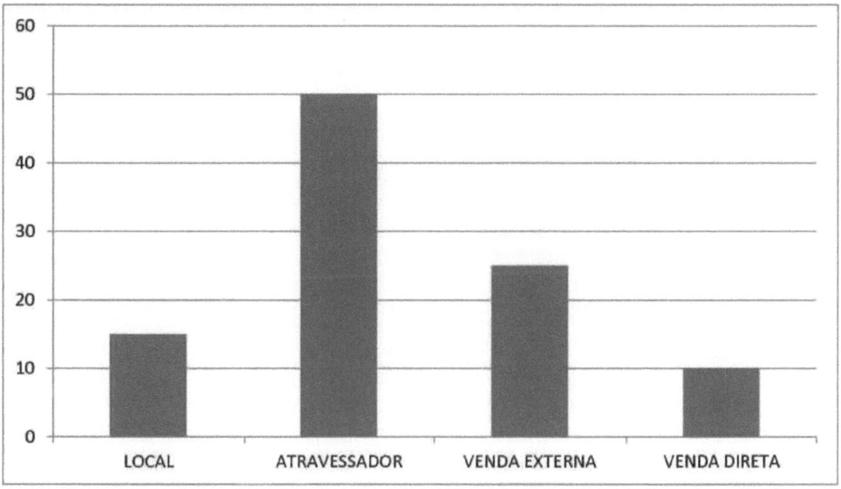

Source: ROCHA, 2018

Most of the bananas produced by the producers surveyed are sold to middlemen, around 50%; 25% of production is sold to other cities and states; only 15% is sold to local businesses; and only the remaining 10% is sold directly to consumers.

Graph 4 - Shows the satisfaction index of producers

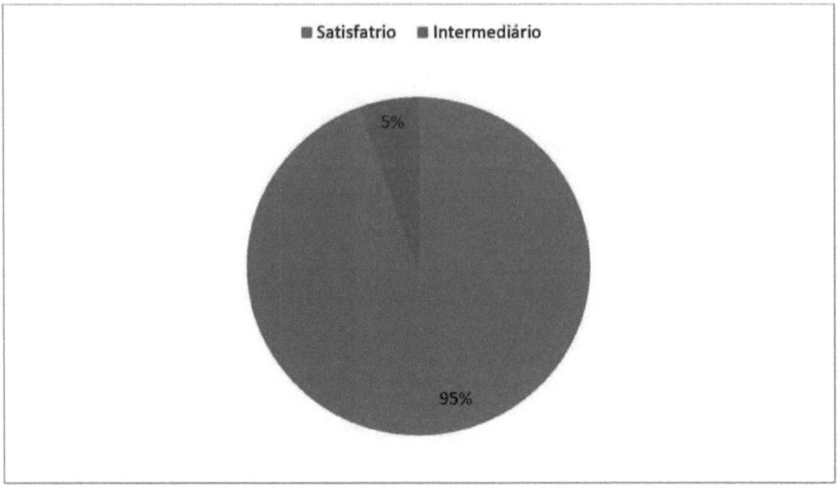

Source: ROCHA, 2018

Based on the interviews with producers, it was possible to observe that 95% of those interviewed are satisfied with their production and 5% are classified as intermediate.

As the graph above shows, planting bananas is very satisfactory, as it is an investment with a quick return, depending on the conditions imposed.

Banana cultivation has become a promising option for producers who wish to diversify their crops. To achieve this, producers need to obtain high yields, which will compensate for the investments and also generate profits. Banana cultivation needs to be carried out properly, always depending on the plant's requirements such as water, light, pest and disease control, as well as all the necessary cultural treatments.

7. CONCLUSION

According to the research carried out, it can be deduced that the socio-economic importance of the banana tree for the Cariri region has become very active, as it contributes greatly to the economy of the Municipality of Missao Velha, both in terms of generating jobs and income, as well as enriching the diet, and today this crop is the main fruit grown by both large, medium and small producers in the region.

8. BIBLIOGRAPHICAL REFERENCES

BORGES, A. L.; SOUZA, L. S.; OLIVEIRA, A. M. G. (Org.). **Fertilizing for high productivity and quality**: tropical fruit trees in Brazil. Fortaleza Agroindùstria Tropica, 2009 (IIP Bulletin, 18). Available at: <http://www.cnpat.embrapa.br/cnpat/down/index.php?pub/FruteirasTropicaisdoBrasil.pdf>. Accessed March 14, 2018.

CARVALHO, J. E. B. **Banana:** phytosanitary. Brasilia, 2000. Available at:<http://www.frutvasf.univasf.edu.br/images/banana1.pdf>. Access

on: June 5, 2018.

CERVO, A. L.; BERVIAN, P. A. **Metodologia cientifica**. 5. ed. Sao Paulo: Pearson Prentice Hall, 2002.

FERREIRA, F. S. et al. Fruit growing in Ceará: evolution and trends in the Cariri metropolitan region. **Enciclopédia Biosfera**, 2014. Available at:

<http://www.conhecer.org.br/enciclop/2014a/AGRARIAS/A%20%20fruticultura%20Cariri.pdf>. Accessed on: March 15, 2018.

FIORAVANÇO, C. J. World banana market: production, trade and Brazilian participation. **Informaçoes Econômicas**, Sâo Paulo, v.33, n.10, p. 16-27, Oct. 2003.

GUEDES, K. F. **Family farming in the sustainable banana production of land credit.** 2016. 24 f. Course Conclusion Work - TCC (Technology in Irrigation and Drainage) - Centec Faculty of Technology - Fatec Cariri, Juazeiro do Norte, 2016).

IBGE (Brazilian Institute of Geography and Statistics). Monthly survey to forecast and monitor agricultural harvests in the calendar year. **Surve. System. Prod. Agric.**, Rio de Janeiro, v.28, n.12, dec. 2014.

INTER-AMERICAN INSTITUTE FOR COOPERATION ON Agriculture (IICA). **Family farming**: rurality, territory and public policy. Brasilia: IICA, 2015 (Sustainable Rural Development Series, v.23).

IPECE (Institute of Research and Economic Strategy of Ceará). **Profile MunicipalMissâo Velha** .2017. Available at:

<http://www.ipece.ce.gov.br/perfil_basico_municipal/2017/Missao_Velha. pdf/>. Accessed on: May 20, 2018.

MARCONI, M. A.; LAKATOS, E.M. **Técnicas de pesquisa**. 6. ed. Sâo Paulo: Atlas, 2006.

NASCIMENTO, L. A.. **Implementation and evaluation of the irrigation system in a banana orchard in Sitio Paraiso Verde, Missâo Nova- CE.** 2010. 34 f. Internship Report (Technology in Irrigation and Drainage) - Centec Faculty of Technology - Fatec Cariri, Juazeiro do Norte, 2010.

NASCIMENTO,V. A. B. **Cultivation, harvesting and post-harvesting of bananas (*Musa sp*) at Frutacor, Apodi - CE.** 2009. 44 f. Internship Report (Technology in Irrigation and Drainage) - Centec College of Technology - Fatec Cariri, Juazeiro do Norte, 2009.

OTERO, M. Foreword by IICA. In: INTER-AMERICAN INSTITUTE FOR AGRICULTURE COOPERATION (IICA). **Family farming**: rurality, territory and public policy. Brasilia: IICA, 2015 (Sustainable Rural Development Series, v.23). p. 17 - 18.

SANTOS, J. R. **Banana marketing process:** a field study with small producers in the municipality of Cacoal-RO. 2014. Available :

<http://www.ri.unir.br/jspui/bitstream/123456789/1010/1/Artigo%20Final %20formatado.pdf>. Accessed on: May 3, 2018.

SEBRAE (Brazilian Micro and Small Business Support Service). Banana: summary. **SEBRAE/ESPM Market Studies**, Sept. 2008. Available at:<http://files.prof-vanderufersa.webnode.com.br/200000038-ad346adaec/Banana%20estudo%20de%20mercado%20SEBRAE.pdf>.

Accessed on: 30 Apr. 2018.

SENA, J. V. C. Aspects of banana production and the market in the Northeast. **Informe Rural ETENE**, Fortaleza, year 5, n. 10, jul. 2011. Available at:

<http://edi.bnb.gov.br/content/aplicacao/etene/etene/docs/ire_ano5_n10. pdf>. Accessed on: May 15, 2018.

SILVA, V. F. V. et al. Potential use of the new herbicide indaziflam in banana orchards. **Revista Brasileira de Herbicidas**, v.16, n.4, p.325-332, Oct./Dec. 2017.

VIDAL, M. F.; XIMENES, L. J. F. Recent behavior of northeastern fruit growing: area, production value and marketing. **Caderno Setorial ETENE**, year 1, n. 2, Oct 2016.

SIMOES, P. A. F. **Applicability of a participatory diagnosis with a view to implementing irrigated banana cultivation (*Musa spp.*) in a land credit production unit** 2011. 38 f. Course Conclusion Paper (Technology in Irrigation and Drainage) - Centec College of Technology - Fatec Cariri, juazeiro do norte, 2011).

WOLF, G. V.; ARIAS, D. Foreword by the World Bank. In: INTER-AMERICAN INSTITUTE FOR AGRICULTURE COOPERATION (IICA). **Family farming**: rurality, territory and public policy. Brasilia: IICA, 2015 (Sustainable Rural Development Series, v.23).

APPENDIX A - INTERVIEW SCRIPT

1) How long have you been producing bananas?

2) Did you produce other things before?

a. [if yes] What? Why did you start planting bananas?

3) Are you satisfied with banana production?

4) Which variety of banana do you grow?

5) How do you get the seedlings?

6) How do you fertilize?

7) How is the harvest done?

8) What pests and diseases attack bananas? How are they controlled?

9) Is your production irrigated [if yes] What irrigation system do you use?

10) How much land have you planted? Are you thinking of expanding this area?

11) Do you know what your production costs?

12) How much do you produce each year?

13) How do you market your produce?

14) Since you've been producing bananas, has your life and that of your family improved in any way?

15) Do you take part in any projects for producers?

16) Do you belong to any producers' associations?

APPENDIX B - PHOTOS OF THE SITIO DA SERRA COMPANY

Cooling chamber

Bagging the banana bunches

Boxes for transporting fruit

Weighing machine

Curl transport handle

Fruit cleaning tank

Conveyor belt for transporting the boxes

Centrifugal pump

APPENDIX C- PHOTOS OF BANANA PRODUCERS

Fertigation tank

Easel

Banana plantation

Banana heart or banana navel.

APPENDIX D - PHOTOS OF THE ASSOCIATION SITIO BARREIRAS

Interview with the association's producers

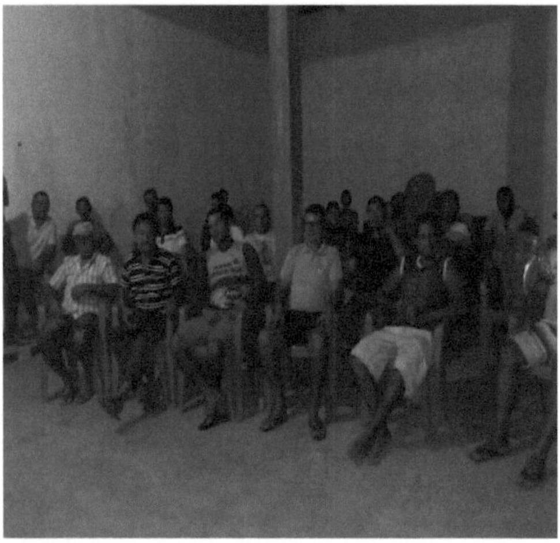

Producers' association

ANNEX A: PHOTOS OF BANANA PESTS AND DISEASES

Figure 1- Leaf with black Sigatoka

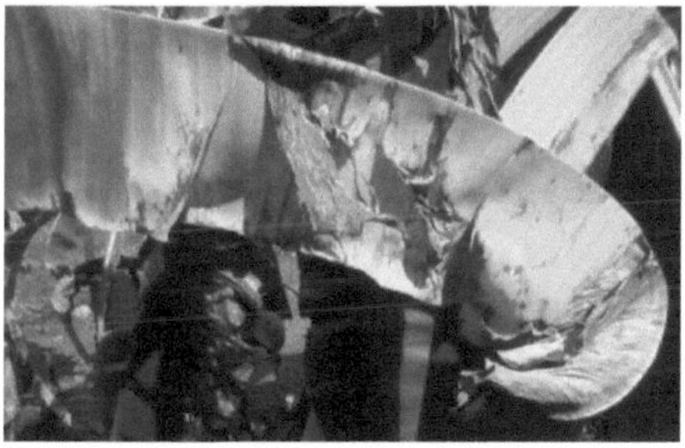

Figure 2 Leaf with yellow Sigatoka-

Figure 3 Rhizome borer adult specimen

Figure 4 Plant killed by nematodes

Figure 5 Fruits with Moko

Figure 6 Banana tree with Panama disease

MIX
Papier aus verantwortungsvollen Quellen
Paper from responsible sources
FSC® C105338

Printed by Books on Demand GmbH, Norderstedt / Germany